Careers in Animal Science

A Guide to Finding Your Dream Job

Molly Nicodemus, PhD

Mississippi State University

Kendall Hunt

publishing company

Cover image courtesy of Molly Nicodemus
All photos and art courtesy of Molly Nicodemus unless otherwise noted

Kendall Hunt
publishing company

www.kendallhunt.com
Send all inquiries to:
4050 Westmark Drive
Dubuque, IA 52004-1840

Copyright © 2017 by Kendall Hunt Publishing Company

ISBN 978-1-5249-4765-1

All rights reserved. No part of this publication may be reproduced,
stored in a retrieval system, or transmitted, in any form or by any means,
electronic, mechanical, photocopying, recording, or otherwise,
without the prior written permission of the copyright owner.

Published in the United States of America

Table of Contents

Preface   v

Acknowledgements   vi

Chapter 1   1
Introduction to the Animal Science Industry:
Getting to Know Where You'll Be Working

Chapter 2   5
What Is an Animal Scientist?

Chapter 3   11
What Do I Want to Be When I Grow Up?

Chapter 4   17
Exploring Your Options

Chapter 5   23
Laying Down the Foundation

Chapter 6   27
Life beyond Undergraduate School

Chapter 7   33
Getting to Know the People in the Industry

Chapter 8   37
The First Step to Your Dream Job: *The Application*

Chapter 9   45
Tips for Presenting Yourself Well in a Resume

Chapter 10   55
Tips for Presenting Yourself Well in a Cover Letter

Chapter 11   61
Tips for Presenting Yourself Well in Person

Chapter 12   67
Where to Go from Here: *Goal Setting*

Preface

The animal science industry is growing and rapidly changing, and with these changes, comes a new crop of students with a diverse background that want to dedicate their lives to animals. This growth in the industry allows for new jobs in new areas that are far different from what was seen in the early 1900's. The traditional student growing up on a family farm is now outnumbered by students with a more unique experience with animals and these animals often don't include the traditional livestock animal. This diversity opens up new jobs in the animal science industry beyond a rancher, farmer, or veterinarian, and this can be both exciting and overwhelming for an undergraduate student in an animal science degree program.

The aim of this workbook is to take the student through a series of steps that introduce a student to the animal science field, and then, have them build the tools needed to get the careers that they might be interested in. Each chapter begins with a short introduction to these steps, followed by a worksheet for the student to work on. Students are encouraged, along the way, to interact with those in the industry, whether this work is with an instructor or an employer. These connections are valuable in making decisions about your career and helping to determine whether you are on the right career path. All of these activities require some footwork by the student, but this process will assist in developing the skills and connections needed to be successful in the animal science industry.

While the workbook is focused on undergraduate students in an animal science degree program, it is never too early, nor too late to go through these steps. For the high school student, these steps may tell you if this is a path you want to undertake in college, and for the older, non-traditional student, these same steps can help fast track what needs to be done if you are looking for a new career path. Even students outside of an animal science degree program, such as those in other agricultural-based degree programs, can find these steps useful in their career planning process.

Acknowledgements

While this workbook is designed for students, the development of this workbook was motivated and guided by my students. The undergraduate and graduate students that I have advised and instructed in my 20-plus years in academia have given me insight into the ever-changing animal science industry and have shed light into some of the challenges that have arisen over the years. Together, we have worked through such issues, and I have been inspired by my students and what they have uncovered about the industry, and about themselves. Even after graduation these same students come back to contribute to my courses giving back the knowledge they learned while working in the industry. Such former students as Ms. Alden Burdine, Mr. TJ Carr, Dr. Patrick First, and Mr. Tyler Herndon have continued to contribute to my courses and guided our future animal scientists. I believe my students have made not only this workbook, but the industry stronger through their passion and drive for the betterment of animals.

Although my background in agriculture communications and technical writing has assisted in educating my students and advisees, the wealth of information on the technical aspect of career planning and tools needed for achieving these careers has come from Mrs. Mary Havens, a retired business communications teacher with over 30 years in the educational system. The further development on the approach taken in this career planning workbook and the associated course was done through the advisement of Dr. Toree Williams, an assistant professor in animal science at University of Findlay. Dr. Williams's own experience taking a career planning course as an undergraduate and through her time as an educator, both in the college-setting and with a younger audience, gave a unique perspective on the steps discussed in this workbook.

To make each step real to those working through the workbook, pictures were utilized by photographers that are a part of the animal science industry: Mrs. Sherri Mitchell, Dr. Kristen Slater, and Mr. Randy Dailey. Their perspective of those in the industry, utilizing the camera lens, tells us much of what the industry is and why we are here. Even the human subjects in the photographs used in the workbook are directly or indirectly involved in the industry: Nick and Bobby Culwell, Spencer Nicodemus, and Dr. Brittany First. Of course, as an animal scientist, I cannot neglect the animal subjects of the photographs as they, like their two-legged friends, remind us of why we are planning a career in animal science, and thus, I give thanks to Val, Aristotle, Dylan, Dude, Blaze, Paris, Socks, and Tex. May their images continue to inspire new animal scientist.

Chapter 1

Introduction to the Animal Science Industry:
Getting to Know Where You'll Be Working

You've always loved animals, and so, of course, you want to go into the animal science industry. A career working with animals sounds perfect for you as your life has revolved around caring and working with animals. However, outside of having animals as a child you know little about the actual industry so your first step is *getting to know where you'll be working*.

The *Animal Science Industry* was first referred to as the *Animal Husbandry Industry*. The practice of animal husbandry has been around for thousands of years, ever since humans first started domesticating animals. Nevertheless, the practice of selectively breeding animals to produce desired traits was not established as a regular practice until the 18th century during the British Agricultural Revolution. In fact, the first systematic selective breeding program for domesticated meat animals was established by a British agriculturalist and scientist named Robert Blakewell. As these practices evolved to include more scientific methodology, the need for more educated individuals in the animal husbandry industry became more evident, and thus, the birth of the animal husbandry degree in the University setting was born.

Timeline: Animal Husbandry Degree Evolution

| Early 1900's Animal Husbandry and Dairy Husbandry degrees available | Prior to 1940's Animal and Dairy combine to Animal and Dairy Husbandry | Around the 1950's Husbandry is dropped and Science is added with programs divided between Animal and Dairy | 1980's Animal and Dairy combined; other programs also combined with poultry and/or veterinary sciences |

As seen in the above timeline, this degree has evolved over time as the demands for the industry has also evolved. With the evolution of the degree, the variety of individuals that have selected such degrees have come from all walks of life, and thus, a potential employer may ask *where do you come from*? *Where do you come from* says a lot about what your background and perspective as it pertains to animal science would consist of. This is your foundation for the start of your academic career, and later, your career goals.

Animal Science Industry Self-Inventory

The animal science industry is widespread in today's society touching many aspects of daily life in about every American's life. Whether you grew up on a farm or in the city, you have been impacted by the animal science industry. From the food you eat to the clothes you wear, you have been affected by the animal science industry.

The objective of this self-inventory is for you to reflect on aspects of your life that may be impacted by the industry or predict where you find you will be impacted by the industry in the future. This self-inventory will give you direction on developing your future career goals as we work through this workbook. In addition, as you go through the self-inventory consider areas in the animal science industry that you are currently limited in so that you can make plans for ways that you can fill in those gaps.

1. What area of the industry do you plan to focus your career (circle one):
    a. Medicine and Health
    b. Nutrition
    c. Breeding
    d. Behavior and Training
    e. Management
    f. Other: _____

2. What animal in the industry do you plan to focus your career (circle one):
    a. Companion animal (dogs, cats, etc.)
    b. Meat animal (cattle, pigs, goat, etc.)
    c. Dairy
    d. Equine
    e. Other: _____

3. What type of animal did you and/or your parents own growing up (circle all that applies):
    a. Dog
    b. Cat
    c. Horse
    d. Cow
    e. Pig
    f. Sheep
    g. Goat
    h. Other: _____

4. What type of setting did you mainly grow up in?
    a. Urban (city or town setting)
    b. Rural (country setting)
    c. Other: _____

5. Circle the youth organizations associated with your career goals that you had participated in as a youth (circle all that applies):
    a. FFA
    b. 4H
    c. Livestock Breed Associations (Name:_____)
    d. Local Youth Club (Name: _____)
    e. Other: _____

6. Basic animal science activities that you participated in, and are competent in performing (circle all that applies):
    a. Management of companion animals (dogs, cats, horses, exotics, etc.)
    b. Management of livestock (beef cattle, dairy, goats, sheep, pigs, etc.)
    c. Training and handling of companion animals (dogs, cats, horses, etc.)
    d. Training and handling of livestock (beef cattle, dairy, goats, sheep, pigs, etc.)
    e. Working farming equipment (tractors, balers, etc.)
    f. Management of farming facilities (building and fixing fences, barns, etc.)
    g. Other: _____

7. After you graduate, your next step will be:
    a. Get a job
    b. Go to Veterinary College
    c. Go to Graduate School
    d. Other Education or Training:_____

8. What course at the collegiate level has been the most useful in your career goals?

9. In looking at your career goals, what is a course or an activity that you would like to participate in before you graduate?

10. If you had to select another major, what would it be and why?

Chapter 2

What Is an Animal Scientist?

Understanding what career options you have available once you graduate begins with understanding what your degree says about you. An animal science degree says to the world that you as a graduate are an *animal scientist*. As a graduate with an animal science degree, the question you must ask is *what is an animal scientist* and *what career options are available for an animal scientist*?

The evolution of the animal science degree from that of an animal husbandry degree found in the early 1900's has expanded the career options for animal science students. The animal science degree, unlike the husbandry degree, opens up the field beyond just production and management of livestock animals.

Animal Science is defined by Wikipedia as *the study of animals that are under the control of mankind*, and thus, an *animal scientist* would be the scientist who works in this field. This definition is quite vague, which can be overwhelming for a student trying to figure out what careers they are qualified for, but on the other hand, this open-ended definition can allow students an open range of careers to select from. Today's animal science students aren't limited to the traditional career path of farming and ranching. The options are endless.

Keeping this in mind, students should begin by finding what definition of animal scientist fits best for them. As students research definitions of what an animal scientist is they will find a variety of definitions. While definitions vary, there will be some *core themes* seen throughout these definitions.

Below note some of the consistent themes that are seen in these definitions. These themes are the core to the curriculum of many animal science degree programs, and thus, a student pursuing a degree in this area should have an interest and background in these areas.

Core Animal Scientist Themes

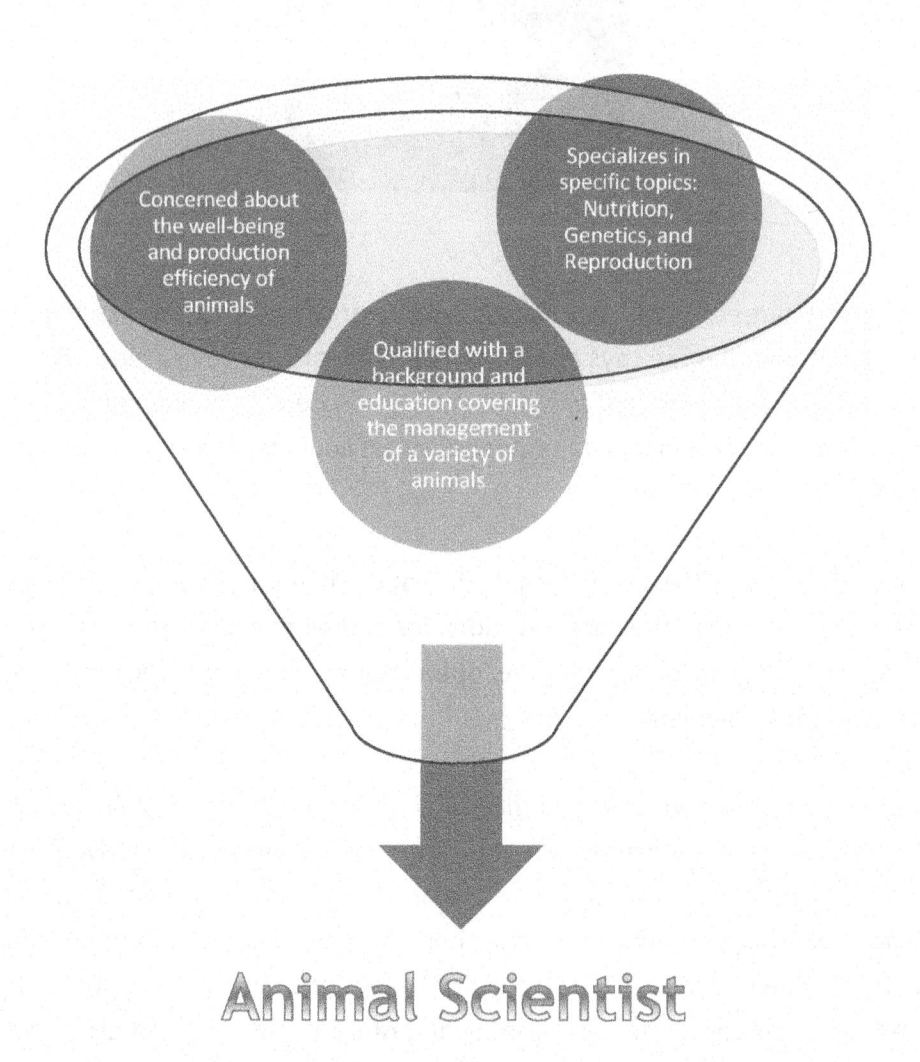

- Concerned about the well-being and production efficiency of animals
- Specializes in specific topics: Nutrition, Genetics, and Reproduction
- Qualified with a background and education covering the management of a variety of animals

→ Animal Scientist

Before you move on to search for a job you need to determine what definition for an animal scientist best describes your expertise to a potential employer. Selection of a definition will tell a student what area of the industry they may be more leaning towards for career selection.

Below are some examples of animal scientist definitions that animal science students have developed during their career planning process. As you prepare for your job search, keep in mind the definition you select and make sure that your application, resume, cover letter, and interview reflects this definition, highlighting all that you have done to make you qualified to be an *animal scientist* as defined by your selected definition.

- *Someone who works to better the welfare of animals, while increasing the efficiency of animal production.*

- *A person who researches all aspects of animals to determine ways to enhance their well-being.*

- *An individual that studies all types of animals from those labeled as livestock to those considered to be exotic.*

- *Someone who studies reproduction, genetics, nutrition, behavior, and production as it relates to a variety of animals.*

- *A person that dedicates their life to the understanding of the production, management, and development of all types of animal species.*

- *An individual that works in various fields ranging from husbandry, production, genetics, diseases, medicine, and sustainability of the animal industry.*

- *Someone who explores topic areas of the animal industry including genetics, nutrition, reproduction, diseases, growth, and development.*

> *A person that conducts research on domesticated animals in the areas of genetics, nutrition, reproduction, diseases, growth, and development.*

> *An individual that studies the biological make-up and production activities of animals that are under the care and management of humans.*

Definition Activity

To begin the process of understanding your degree and what it says about you, you must first define what it is that you'll be once you finish your degree. An animal science degree means you are an animal scientist, and thus, the question is *what is an animal scientist*? Begin first by researching the word *animal scientist*, and then, develop your own definition.

Below in the chart students are to find three definitions published by other individuals besides those used in this workbook. Students are to write down the definitions and include where the definitions came from.

Location of the Definition	Definition

From the definitions above, students are to develop their own definition. The definition should cover what the student wants to focus on in their career selection.

Student Definition:

_____

_____

_____

Chapter 3

What Do I Want to Be When I Grow Up?

Photo courtesy of Dr. Kristen Slater

Identifying careers to fit your degree begins first with defining what your degree says about your qualifications. Once students have defined what an animal scientist is to them, the next step is searching what jobs match that definition. Students should keep that definition in mind as they search for jobs. While most students enter an animal science degree program to become a veterinarian, students should adventure beyond this traditional career choice looking at other fields in the industry. Students may be surprised at what options are available.

Get creative. Career options are necessary as the first career choice may not always be the best choice for you. Furthermore, it is helpful, once a list of jobs are acquired, to rank those jobs according to desirability as this gives students a direction as to where to focus their career path as they navigate through their undergraduate degree. This also gives students a place to start in the job search, which is not a process that happens overnight.

Below is the list of the animal scientist definitions that undergraduate students have developed, and underneath each definition are jobs that students selected that fit the definition they developed. Note the majority of jobs listed, despite the differences in these definitions, are jobs requiring a doctorate in veterinary medicine. While the field of veterinary medicine is popular for most undergraduate students in an animal science degree program, students need to question whether there are enough jobs available for all students interested in a veterinary career. For those students leaning towards a veterinary degree, look at some of the other jobs listed and keep your mind open to alternative career options. College is about having options and today's animal science degree allows for such options as seen below.

Animal Scientist Definitions & Associated Careers

Definition: *Individual that applies the biological, physical, and social sciences to the issues associated with livestock husbandry.* Careers:
- USDA Inspector
- Artificial Insemination (AI) Breeding Technician
- Livestock Production Manager
- Livestock Feedlot Operator
- Sire Program Consultant

Definition: *Someone who studies the nature and well-being of animals to better manage animals and to address food safety concerns for humans.* Careers:
- Livestock Veterinarian
- Meat Research Scientist
- Animal Industry Lawyer
- FDA Veterinarian
- Animal Welfare Educator

Definition: *Individual that applies their knowledge of various types of animals to assist with the diagnosis and treatment of these animals and to more effectively educate the public and producers.* Careers:
- State Extension Specialist
- Animal Nutritionist Veterinarian
- Beef Industry Lobbyist
- Animal Behaviorist
- Veterinary Pharmacologist

Definition: *Person that strives for the betterment of all types of animals through the study of production and management methods.* Careers:
- Veterinary Anesthesia Technician
- Animal Science Professor
- Equine Reproduction Veterinarian
- Mixed Animal Veterinarian
- Animal Health Product Sales Representative

Definition: *Person who studies every component of the animal in order to better the quality of animal's life, as well as to benefit people and the environment as it relates to these animals.* Careers:
- Equine Veterinarian
- Animal Geneticist
- Livestock Insurance Agent
- Animal Industry Public Relations Consultant
- Environmental Risk Manager

Definition: *Someone who focuses on the basic science of production and management of both domesticated and wild animals.* Careers:
- High School Agriculture Teacher
- Breed Analyst
- Animal Shelter Manager
- County Extension Agent
- Fish and Game Warden

Definition: *One who studies the behavior, reproduction, anatomy, and physiology of all types of animals to improve the health and welfare of these animals.* Careers:
- Exotic Animal Veterinarian
- Animal Nutrition Sales Representative
- Large Animal Veterinary Clinical Instructor
- Marine Biologist
- Animal Reproductive Researcher

Definition: *Someone who studies management practices of animals for increasing production output including such individuals as nutritionists, veterinarians, animal behaviorists, and producers/breeders.* Careers:
- Small Animal Veterinary Surgeon
- Veterinary Pharmaceutical Sales Representative

- Exotic Animal Researcher
- Marine Animal Veterinarian
- Animal Trainer

Definition: *Someone who observes and studies domesticated animals that are used for food production.* Careers:
- Veterinary Technician
- Swine Production Manager
- Feed Mill Processing Manager
- State Inspector for Dairy
- Animal Nutritionist Consultant

Definition: *Individuals that study the biological aspects of domesticated animals and their respective management activities in order to benefit the lives of both animals and humans.* Careers:
- Equine Farrier
- Veterinary Clinic Manager
- Animal Welfare Specialist
- Equine Rescue Specialist
- Animal Pathologist

Definition: *Someone who has dedicated their life to the improvement and maintenance of the well-being of animals.* Careers:
- Service Dog Trainer
- Wildlife Rehabilitation Center Veterinarian
- Companion Animal Therapist
- Animal Rescue Officer
- Racetrack Facility Manager

Definition: *A person who develops more efficient and effective methods of producing and processing meat, poultry, eggs, and milk.* Careers:
- Seedstock Producer
- Animal Sales Consultant
- Dairy Animal Nutritionist
- Beef Cow-Calf Manager
- Waste Management Consultant

Job Shopping Activity

You've started down an animal science degree, and now, have defined what that degree means to you. At this time, a student needs to begin shopping for what jobs interest them and fit their degree. Knowing what career options you are working towards as you navigate through your degree program assists in motivating students giving them a goal to work towards and a *foundation* in the career planning process. The animal science degree opens many career options, and that can be overwhelming to students unless they begin to narrow down their path to making a Plan A, Plan B, ... Plan E as it relates to career options. Options are good as long as they are focused.

To achieve this career planning *foundation*, students need to find multiple jobs that peak their interest and write them down. Writing these options down makes the job search seem more formal and gives visual confirmation of what you are working towards. In this exercise, students should start first by writing above the chart given below their animal scientist definition as this is a reminder of what they think they are qualified to do once they complete their degree program. Next, students are to list 5 job options in the chart that fit their animal science definition and rank these options with their top choice being listed as first. Next to each job, give a simple one sentence definition of that job.

Animal Science Definition:

Job Choices	Job Description
1.	
2.	
3.	
4.	
5.	

Chapter 4

Exploring Your Options

As students make their selection of what they want to do with their degree and what career they potentially want for their future, they next need to thoroughly research their career options. This process will assist in ensuring your job selection meets your interests.

Researching jobs consists of answering some of the common questions most undergraduate students ask as they navigate through the job selection process. These questions answer what type of lifestyle to expect once you are in this career. Again, each student will have their own idea of what they want for their future including how they want to live and the type of lifestyle they hope to have. At this point, students need to have a serious conversation with themselves asking such specific questions as what amount of family time or disposable income do I want?

Below are some examples of questions undergraduate students are concerned about when selecting a career path. These questions are for any student, not just those in an animal science degree program. The answers to these questions and others like it should indicate to a student if this career option fits what a student wants to do for the rest of their life.

Career Questions

- Is this job labor intensive?
- Does this job take place primarily in an office?
- Will I work weekends on a regular basis?
- Will I work evenings on a regular basis?
- Does this job require extensive travel?
- Can I be my own boss?
- Does this job have the potential for a high paying salary?
- Will my income be based on my performance?
- Do I need a degree beyond just an undergraduate degree?

Where do I begin to answer these questions? Obviously, most undergraduate students will say the internet. This is a good place to begin, although students need to be careful and use trustworthy websites so that they don't base their job selection on false or bias information. While researching on the internet, students can locate individuals working in these careers. Interviewing people working in the careers you are interested in can assist in better understanding your potential career.

Keep in mind that each person has their own perspective in that what you consider labor intensive may not be so to someone else. This is why the next recommendation is to walk in the shoes of these professionals by shadowing them to find out what a day in the life of a professional in your career choice consists of. We will address these steps in the following chapter.

Below are some of the answers to the above questions that undergraduate students have researched and answered through internet searches, interviews, and internships.

Answers to Questions:

Labor intensive jobs:
- Farm manager
- Veterinarian
- Veterinary technician
- USDA inspector

Office jobs:
- Animal industry lawyer
- Beef industry lobbyists
- Animal science professor
- Animal geneticist

Jobs with irregular working hours (weekends and/or evenings):
- Farm manager
- Livestock consultant
- Veterinarian
- Veterinary technician
- Extension agents (county and state)

Jobs requiring extensive travel:
- Artificial insemination (AI) technician
- Beef industry lobbyists
- Animal products sales representative
- USDA inspector

Jobs where you're your own boss:
- AI technician
- Farm owner
- Veterinarian
- Livestock consultant

Jobs with a potential for a high salary:
- Livestock consultant
- Nutritionist
- Veterinarian
- Animal geneticist
- Animal industry lawyer

Jobs where income is based on performance:
- AI technician
- Animal industry lawyer
- Beef industry lobbyist
- Veterinarian
- Animal products sales representative

Jobs typically requiring a degree or certification beyond a bachelor's degree:
- AI technician
- Veterinarian
- Animal industry lawyer
- Animal science professor

Career Comparison Activity

Once a student determines what careers are of interest to them, the next step is to research those careers so that comparisons can be made and assist in deciding if that career is right for you. In this worksheet, from the career list that students have put together in the previous worksheet, students are to compare their selected five careers by answering the questions below. Start by listing below the five careers selected from the previous worksheet, and then, begin researching each of the listed careers to answer the following questions.

Five careers selected: 1. _____ 2. _____
3._____ 4._____ 5. _____

1. Which career seems to be the most physically, labor intensive?

2. Which career mostly takes place in an office?

3. Which career requires working weekends on a regular basis?

4. Which career requires working evenings on a regular basis?

5. Which career requires extensive travel?

6. Which career allows you to be your own boss?

7. Which career requires you to be a team player?

8. Which career gets paid the most (include amount)?

9. Which career gets paid the least (include amount)?

10. Which career is your income based on your performance?

11. Which career requires a degree beyond a bachelors degree (list the degree)?

12. Which career does not require a bachelors degree?

13. Which career requires a specialized certification or specialized training (list the certification)?

Chapter 5

Laying Down the Foundation

The word "resume" comes from the French word résumé meaning "summarized". A resume is a document that assists potential employees in showing off what they have to offer a potential employer. Resumes should highlight education, experience, and other relevant activities.

As students go through their academic career, they should determine what future career they may be interested in pursuing and what experiences and education will get them to their career goal. This should be done before a student graduates with students evaluating regularly what activities may help to build their resume to attract potential employers. This experience does not have to be limited to the right degree or relevant previous jobs, but can include more creative approaches that demonstrate to a potential employer that you have the right background to be a successful employee such as volunteer work or project experience.

Below are suggestions from undergraduate students working on an animal science degree for developing a resume that reflects the potential of being a successful employee.

Resume Development Activities

- First and foremost for most potential employers is obtaining a degree that matches the requirements of the selected career. Most recommend degrees above just a bachelor's degree, although pursuing a minor during your undergraduate career can assist in opening up career options. In researching careers, most undergraduate students reported that the careers selected required specific degrees, if not an advanced degree, so early research, prior to a student's junior/senior year, is needed concerning career requirements.
- Another recommendation is research and/or internship activities that build upon experience if students are lacking previous employment that relates specifically to the career selected. Current degree programs can offer students class credit for research and/or internship activities. Information about these programs is available through academic advisors and career counselors.
- Some students may take courses that complement their degree including courses in public speaking, writing/journalism, economics, and marketing. While these courses may only be classified as an elective, these electives can showcase a unique background from other applicants with similar degrees.
- Additional suggestions from potential employers for resume building include volunteering, apprenticing, certifications, study abroad, seminars, club/association memberships, and entry-level jobs related to the future career. In the end, students should research individuals that are currently in their potential careers to see which of these activities assisted in reaching their career goals.

Through the activities and experiences given above students will be able to make the connections with individuals that can offer references needed for obtaining a job. Reference development is emphasized by potential employers, and although references are not commonly listed directly on a resume, they are an essential component to career obtainment and something that needs to be worked on prior to graduation.

Resume Building Activity

No matter how much time you put into designing the look of your resume, without the right background and experience, the format and design of the resume will not help in getting the career you desire. This background and experience takes time to develop so the earlier you can start, the better.

First, start by listing what things you have done that may be desirable for a potential employer, and then, follow up with items you may need to pursue in order to acquire your dream job. Below is a worksheet to help students visualize areas they may be lacking on their current resume.

Jobs of interest (list the 5 jobs of interest selected in the previous worksheet):

1.

2.

3.

4.

5.

List items on your current resume that would be of interest to potential employers in the jobs listed above (list the top 5 items you have on your resume):

1.

2.

3.

4.

5.

From your research of these jobs what additional items would you need to experience or participate in during your undergraduate career that would help make your resume more attractive to potential employers in the jobs listed above (list top 5 items giving specific examples)?

1.

2.

3.

4.

5.

Chapter 6

Life beyond Undergraduate School

As students prepare to graduate with their bachelor's degree, they need to have in mind their career options, and this becomes particularly important for those careers that require more advanced degrees. Once students decide to go down a career path that requires a graduate degree, they must begin looking for options for graduate school including researching the degrees that meet the needs of the selected career path, schools that offer these degrees, and the requirements for getting accepted into these degree programs. Students should interact with those faculty members associated with these degree programs to find a potential advisor and to learn more about specifics of the program.

This research needs to be done early to meet potential application deadlines. For most graduate programs, applications are recommended to be turned in no later than the semester before starting graduate school, and thus, if a student wants to start graduate school in the fall semester, they should apply by the start of the prior spring semester. This is particularly important if that potential graduate student is looking for some type of funding from the potential graduate program such as a graduate assistantship/stipend.

Below are some questions undergraduate students should research as they shop for a graduate program. As you read through the questions below, you will also see some of the answers undergraduate students in an animal science program have found through their research. These answers should help to get a student started in researching programs on their own, but no matter what, students should take the time to research these questions for any degree program they may be considering. Keep in mind that degree programs can change overtime so the research needs to be current and up to date. Contact directly faculty and staff associated with these programs and take time to personally meet with them.

Graduate Program Questions & Answers

What are my graduate degree options?

Many graduate programs may not be specifically labeled as an animal science graduate degree. For example, Mississippi State University offers a Master of Science in Agriculture with a concentration in Animal and Dairy Sciences and a PhD in Agricultural Science with a concentration in Animal and Dairy Sciences.

Some programs may also offer more specific areas within the animal science discipline. For example, additional interdisciplinary graduate programs are available through the Animal and Dairy Sciences Department in conjunction with other departments (Biochemistry, Poultry Science, etc.) at Mississippi State University including concentrations in Genetics, Animal Nutrition, and Animal Physiology.

While the majority of undergraduate students focus on graduate degrees that are directly related to animal science, some animal science students will pursue graduate degrees in other areas such as Biological Sciences, Biochemistry, Agribusiness, and General Agriculture. These students can tie in their animal science background and interests by taking animal science-related courses during their graduate career, select an animal science faculty member to be on their graduate committee, and focus on a thesis research project that has a subject matter that is directly or indirectly-related to animal science. Some master's degree programs will allow students to do a minor in which animal science students pursuing a degree outside of animal science can select an animal science minor to help continue their work in animal science.

Lastly, masters programs can be divided into thesis and non-thesis degrees meaning you can select a research-based degree program or stick to more of a course-specific degree program. Thesis research can be timely, but a thesis-based graduate program will allow for a student to go on to pursue a PhD.

Where to complete these graduate degrees?

To strengthen a student's resume, students are encouraged to look at graduate programs besides those Universities where they did their undergraduate work. Most undergraduate students will still select Universities that are closer to home. The Southeast Region, for example, has a wide variety of Universities offering such programs, while allowing for a student to still stay fairly close within the region. Examples include Texas A&M, Louisiana State University, University of Florida, Auburn, Georgia State University, University of Georgia, University of Tennessee-Knoxville, Florida State University, Clemson, and North Carolina State University.

If students venture outside of their region, usually they will do so to search out some of the bigger, well-known animal science programs such as Kansas State University, Purdue University, The Ohio State University, and Michigan State University. All of these universities and other Land Grant Universities offer both MS and PhD graduate programs with the majority having graduate degrees associated specifically with animal science. In the end, today's animal science student has many options for selecting a University for pursuing their animal science graduate degree, and thus, should take advantage of these options.

How do I get into these graduate programs?

Majority of animal science undergraduate students report not meeting the minimum requirements for the graduate programs they are considering. While graduate school requirements are important, if you don't find a faculty advisor, even the best grades and scores won't help for you getting into a graduate program. Thus, the first step is finding an advisor that will accept a student as their advisee. Keep in mind, these openings are often limited by what funding an advisor has available.

Nevertheless, there are basic requirements for entrance to graduate programs with the most common requirements including a minimum GPA, usually set at a 3.0, and a GRE, that is commonly set at scores at or above the 20th percentile in each of the 3 areas: verbal, quantitative, and analytical writing. While most graduate programs require a minimum GPA of 3.0, all offer provisional admissions for those students under a 3.0 GPA. This type of admissions will require the student to meet the 3.0 GPA requirement within the first semester of their graduate work.

Many of the graduate programs in animal science don't publish a specific GRE score, and several, only check to make sure you took the GRE, not what score you got. Most graduate programs will require certain courses taken as an undergraduate such as animal breeding, animal nutrition, animal reproduction, animal anatomy and physiology, and an animal production species-specific course.

Finally, students are recommended once they find a potential advisor to discuss additional requirements before applying to a program such as research experience or additional coursework. The earlier a student begins the communication process with a potential advisor the more time they will have to meet those requirements and any additional recommendations. This communication process can begin as early as a student's junior year of undergraduate school.

Graduate Program Activity

As students prepare to graduate with their bachelor's degree, they need to have in mind their career options and particularly those careers that require more advanced degrees. Once students decide to go down a career path that requires a graduate degree, they must begin looking for options for graduate school including researching the degrees that meet the needs of the selected career path, schools that offer these degrees, and the requirements for getting accepted into these degree programs. Students should interact with those faculty members associated with these degree programs to find a potential advisor and to learn more about specifics associated with the program. This research needs to be done early to meet potential application deadlines.

Below are questions students should answer in researching advanced degree programs. Students should look at a minimum of three degree programs including at least one outside of the state. List below what Universities were selected. Then, answer the questions below from the three Universities selected.

Selected Universities (list 3 examples): 1. _____

2. _____ 3. _____

Questions:

1. From the Universities selected, what graduate degree programs were found besides animal science that would be of interest to you to pursue (give at least one example)?

2. Besides Universities within your home state, what other colleges and/or universities offer graduate programs in animal science or closely-related fields (give at least one example)?

3. Which Universities listed specific minimum GPA requirements and what were these requirements?

4. Which Universities listed specific minimum GRE requirements and what were these requirements?

5. Which Universities listed specific coursework that needed to be completed prior to the start of the graduate program and what were these required courses?

6. Which Universities listed specific coursework that needed to be completed during the graduate program and what were these required courses?

Chapter 7

Getting to Know the People in the Industry

Photo courtesy of Randy Dailey

As mentioned in the previous chapter, the people of the industry are critical when you are applying for a job as references are a part of the job application process. References don't happen overnight as most want enough experience with that individual to comfortably say that you are appropriate for the job. While family members and friends are tempting as they have had the time with you, these individuals don't say that you have taken the time to build experience within the industry.

Follow what Mr. Rogers use to teach us in that "Who are the people in your neighborhood?". Building a relationship with people in the industry assists in their ability to make a recommendation for you, but it also gives you an opportunity to learn about the industry. Is this an environment you want to spend the majority of your time in? Getting a firsthand look into the career you might select for your future is priceless. This experience can tell you if your career selection is right for you and can be done via internships, volunteer work, and/or shadowing opportunities. Since each perspective is unique, don't stop with just working with one individual as their approach, environment, and duties may be different from others.

Where do I begin? Most undergraduate students are hesitant to ask someone if they have an internship, volunteer opportunity, or are willing to allow you to shadow them. The easiest place to begin is with a family friend that is working in the industry or let's say the veterinarian that your family has used to treat your animals. Direct or indirect relationships will put you at ease on asking for this working opportunity. Even if you know the individual, don't take the asking process causally. Dress as if you are doing an interview and be prepared with your resume. This may seem over the top, but usually those that have known you for a while won't think of you in that capacity. You need to show them you are serious and are ready to be a professional in the industry.

If you don't have those direct or indirect relationships, visit with your professors at your college. Most Animal Science Departments require some form of internship so these professors are already prepared with internship opportunities. These opportunities may be paid or unpaid and may be close to home or out-of-state. Decide before you begin your search if you can do an unpaid internship and have the finances to live out-of-state temporarily. Remember the question is *can you*, NOT *do you want to*. Usually, the unpaid internships are ones where they are so valuable in experience the employer doesn't have to offer money, and of course, out-of-state internships will open up your career options so that you aren't stuck when you graduate with just applying for jobs in state.

While an internship, volunteer position, or shadowing opportunity is not the same as a full-time position you would get after graduation, many of these positions can evolve into full-time positions if you do a good job. Keeping that in mind, your approach to applying for these positions should still be the same as if you are applying for a full-time position. At a minimum, dress nice, have a resume prepared, and be ready for a short interview. These items will be discussed in more detail in the upcoming chapters.

Getting to Know the Industry Activity

Students are to select a career and find an individual to interview using the questionnaire below that works in this career. Keep in mind that during this interview process this may open up an opportunity to discuss potential internships, volunteer work, or shadowing opportunities with the individual that you are interviewing so be professional. This questionnaire needs to also include a signature from the individual interviewed along with the date of the interview.

Identify the following:

Name-

Career Title-

Location of Career-

Questions:

1. How long have you been working in this career?

2. While working in this career, have you always been at your current location, and if not, where else have you been located?

3. Do you plan to stay at this location till retirement or do you plan to eventually move to another location, and if so, where and why?

4. Why did you decide to pick the career that you are now doing?

5. What was the one thing that you did that most helped you to get you into the career you are in today?

6. What is one thing that you would have added to your academic career and associated activities to make it easier for you to get you in the career you are in today?

7. What is the one thing that you would have not done during your academic career that may have helped you in getting you into your career?

8. List two pros and two cons about your career.

9. What is one thing you wish someone had told you about your career before you had started down your career path?

10. If you were to pick another career, what would it be and why?

11. Do you plan to stay in this career until retirement, and if not, why and what tentative plans do you have?

Chapter 8

The First Step to Your Dream Job:
The Application

While the application form may not be a requirement for all jobs that students apply for once they get out of college, the understanding of how to fill out an application in an effective and professional manner is needed before heading out into the job market. Many of the rules to developing a resume or cover letter also applies to an application (correct spelling, proper grammar, truthful and complete information, etc.).

In all cases, students should try to attempt to have other individuals read over their application before turning the application in. The goal is to take your time and be as thorough as possible. Below are some tips for the application.

Application Tips

Hand-Writing Applications:

Although the application process today is usually done via computer, hand-written applications may still be a possibility for some jobs. Always use a pen. Make sure that your hand-writing is neat and legible and that any corrections should be neat using a single strike-through, erasable ink, or white-out. Use the same type of ink and color of ink, preferably black, throughout the application.

Utilizing Space:

Information should fit into the space given. With applications that are electronic watch that information beyond the capacity of the space given may get hidden or cut-off. In a hand-written application, avoid writing outside of the space given.

To avoid not having enough room on the application, make sure that you are not repeating. This includes such things as writing down work activities both under "special skills/training" and under "description of work" in the employment section.

Providing supplemental attachments such as a cover letter and resume will allow for applicants to include information that cannot fit on an application. In the following chapters, development of cover letters and resumes will be discussed.

Blank or Incomplete Sections:

Avoid areas left blank or incomplete. One of the most common areas where items are left blank or incomplete is in the reference section. This lack of information says to the potential employer that either the applicant does not completely know the person giving the reference or the applicant does not want the potential employer to know specifics about the reference person.

Additional areas left blank in other aspects of the application may suggest an applicant that is not really interested in the position, rushed through filling out the application, or lacks the experience needed for the position. Applicants unsure of how to answer a question such as "desired salary" can use terminology that allows for further discussion (Example: "open" or "negotiable") if you get an interview. Just be prepared to discuss these answers.

Complete Names:

Make sure to use proper and complete names for degrees, Universities/Colleges, reference names and their titles/employers, position and company of former employment, and position applicant is applying for. Incorrect or incomplete information may suggest a rushed application, an applicant that is unorganized and not detailed-orient, or untruthful information.

With limited space applicants may try to abbreviate these names/titles, but if possible, try to avoid abbreviations. If the applicant must abbreviate, include the most commonly used and most well-known abbreviation.

If the name/title is used in multiple areas of the application, try to spell out the name/title in the first area that the name/title is given in the application followed by the abbreviation in the following areas. Keep the abbreviation consistent throughout the application.

References:

References are an important part of the application process. Select references wisely. Avoid using family and friends. Try to use individuals from different businesses, and for recent graduates, try to use one instructor, particularly an advisor.

Make sure that you have contacted the individual that you are using for your reference to make sure that they are willing and enthusiastic to do a recommendation. Also, make sure that you have the most current and complete contact information for your references. Give the required number of references as indicated on the application.

Thorough Information:

Make sure you are giving the potential employer the information they are looking for. Read through slowly what each section is asking for and make sure you are interpreting what the section is requesting.

Sections labeled as "special skills and/or training" are the most common areas where these problems occur. These sections are not looking for the applicant's degree, organizations they are involved in, or personal qualities ("hard-worker", "team player", etc.), but rather, details about job-related skills the applicant has developed or training they have been given that is above and beyond what is listed in the previous employment section.

This information should be specific and detailed so that it is not simply "computer skills", but the specific names of the programs they have training or experience working with. This information should also be specific in the amount of training avoiding using vague terms such as "some".

Many times applicants will leave these sections blank as they don't understand what is being asked and/or don't recognize what things they have done that could fit into these sections. A career counselor or an academic advisor can explain to you what is needed in these sections and what experience you have that may apply. These individuals can also make suggestions on how to write up information so that it can apply to these sections such as club-related activities, class projects, and/or volunteer activities.

In any case, these sections can be valuable to the applicant if used correctly and should not be left blank. On the other hand, if a section is left blank use "NA" (*Not Applicable*), to tell the employer you didn't miss that section.

Practice Application

The application is not always a requirement for some jobs, but the application is a good introduction for the applicant when it is used. Most applications are today done online, but some businesses still use a hard copy application. In those cases, you may not always have the option to take home and type it up, and of course, who today has a type writer? This is when writing neatly is essential and taking the time to think about each answer is important. Even if the application is online, it is good to practice first by printing one out to work on your answers before you submit one electronically.

This worksheet is an opportunity to practice this approach. The questions listed in this worksheet are standard for an application. Select a potential position that you might be interested in applying for and keep this position in mind as you fill out the practice application. Blank questions can indicate to a potential employer that you didn't take the time to fill out the application so make sure to fill out each blank as thorough as possible. For questions that you don't have the background or experience in, don't leave blank, but instead, answer with "NA".

Application Questions:

Personal Information

Name:	Address:
Phone Number:	Email:

Employment Desired

Position:	
Desired Salary:	Date You Can Start:

Education History

High School (Name and Location):	
Year Graduated:	GPA:
Subjects Studied:	

College (Name and Location):	
Year Graduated:	GPA:
Subjects Studied:	

College (Name and Location):	
Year Graduated:	GPA:
Subjects Studied:	

General Information

Special Subject of Study and/or Research Work:
Special Training, Certification, and/or Licenses:

Special Skills, Languages, etc.:

Former Employer

Place of Employment (Name and Location):

Position Title: | Years Employed:

Duties:

Place of Employment (Name and Location):

Position Title: | Years Employed:

Duties:

Place of Employment (Name and Location):

Position Title: | Years Employed:

Duties:

References

Name:	
Phone:	Email:
Address:	

Name:	
Phone:	Email:
Address:	

Name:	
Phone:	Email:
Address:	

Chapter 9

Tips for Presenting Yourself Well in a Resume

It's time to introduce yourself to your potential employer. Resumes are a snapshot into your professional life. You want your resume to reflect all that you have done that would make you attractive to a potential employer.

Even the format and design of the resume hints to a potential employer your creativity, whether you work "outside of the box", the time you take in a project, how detail oriented you are, and what you think is important in your professional life. There are no right or wrong answers when it comes to your resume, other than avoiding poor grammar and misspelling. Each resume should be unique and geared towards your personality and professional experience.

With that being said, there are some basic suggestions concerning resumes that people should keep in mind when designing a resume. While creativity is important, most potential employers are looking for certain aspects of a resume so that if you go too far outside of the box you may lose their attention. For example, an undergraduate student that has spent their four years in college actively involved in many programs has every right to go past the traditional one page resume that most potential employers are looking for, but again, you need to tread lightly as too much irrelevant information can deter a potential employer.

Usual rule of thumb is 10 years per page so if you work over 10 years you have the right to move past that second page. If you do go into a second page both pages need to have the same heading and format and the second page needs to at least be ½ page in length.

The best way to judge if you're style of resume would be attractive to a potential employer is to have it evaluated by multiple individuals with multiple backgrounds. Of course, take their input with a grain of salt, meaning use what suggestions you feel fits your personality, as in the end, the resume needs to reflect you and not someone else.

Keep in mind that each resume should fit the job you are applying for so as you accumulate more experience some activities will be left off if they are not relevant for the job you are applying for. Therefore, an evaluator that has relevant experience for the job you are applying for can give you an honest opinion as to what information should be kept on your resume and which information isn't needed. Below are some suggestions for developing your resume along with a worksheet to help in this evaluation process.

Dos and Don'ts of Resume Writing

The following are the Dos and Don'ts of Resume Writing for college students. Students are recommended to look at examples of resumes and cover letters and to have multiple individuals review over their resume and cover letter before giving them to potential employers.

The style of the resume and cover letter should reflect the style of the student so that it can reflect to the potential employer the type of employee that they may be employing.

To *Do* list:

- Be honest. Remember potential employers can learn a lot about an applicant through social media. A small "white lie" on a resume can easily be uncovered researching the internet.
- Keep your format simple enough that it is easy to follow. Start with a resume template and work to make it your own.
- Be creative on your wording so that you don't start to sound repetitive and do highlight information that you want to standout to the potential employer.
- Make sure verb tense matches when you have done a job or participated in the activity so that past jobs have past tense used in the description.
- Spellcheck and have others spellcheck as even the computer can make mistakes.

- Make sure you include at the top of the page a header with your name and your full, current contact information: Name, Address, Phone Number, and Email Address.
- Avoid going too small on font, 10-12 points suggested, so that it is easier to read and use more standard style fonts, Arial, Calibri, Garamond, Helvetica, Times, or Tahoma.
- Make sure your name stands out with the largest font and even a font that stands out from the rest.
- Make headings and sub-headings stick out with bold and full caps.
- Utilize your space so nothing larger than 1" margins.
- Resume paper makes the resume look more professional and helps to avoid wrinkles as it is a heavier paper.
- Don't go too creative on color of paper so stick with soft cream color or a pale gray.
- Bullets help to keep the information organized and less cluttered. The indent for each bullet should help for items to standout, but shouldn't indent too much that you are wasting space.
- Be specific giving numbers and years such as number of years volunteered or date you started and ended a job.
- Use facts and figures keeping the information exact and to the point avoiding embellishments and this should include specific names of positions, locations, certifications, etc.
- Use volunteer work, class projects, and club activities if it is relevant to the job and fills in where job experience may be lacking.

One of the most common questions students have is *where do I put information about references*? Instead of having "references upon request" on your resume, just include a separate page that lists your references with their full addresses. Number of references usually requested by employers is three, but ask first before turning in your resume. Make sure this reference page has the same heading, formatting, and paper as the resume.

The following page is an example of a well-formatted resume for a college student that includes an example of a reference page. Along with reviewing over the resume in this chapter, take time on the internet to look at other examples, and then, develop something that matches your interest and background.

TERRI LYNN SMITH
240 Scenic Drive, Saint Charles, Missouri 63320
terri129@missouristate.edu
310-310-1199

OBJECTIVE
To work with a research institution that provides quality education and utilizes my best teaching skills for students of natural science.

EDUCATION
M.S. Natural and Applied Science, Agriculture Education
Missouri State University, Springfield, Missouri / August 2009-December 2011 GPA 4.0
- Developed a full semester animal science curriculum suitable for third and fourth year secondary students and for 100 level college students
- Obtained a State of Missouri Agriculture Education Secondary Teacher's Certificate
- Assisted in equine and small ruminant research studies centered on nutrition, metabolism and reproduction

B.S. Animal Science with Minors in Agriculture and Equine Science
Missouri State University, Springfield, Missouri / August 2005-May 2009 GPA 4.0
- Recipient of the Darr School of Agriculture Ambassador Scholarship

TEACHING EXPERIENCE
Teacher's Assistant
Missouri State University, Springfield, Missouri / August 2009-December 2011
- Assisted primary instructor with teaching Equestrian Science, Introduction to Riding, Riding for Training, Problems in Animal Science
- Developed and presented lectures on animal physiology, nutrition, genetics, soil science, and horticulture
- Tutored students in Biology, Soil Science, Animal Physiology, Animal Nutrition and Reproduction
- Coached beginning to advanced collegiate athletes
- Supervised students in safe riding and handling practices

Marshfield High School Student Teacher
Marshfield R-1 School District, Marshfield, Missouri / August 2011-December 2011
- Developed lesson plans and taught Animal Science, Plant Breeding, Plant Genetics and Science, Agriculture Business, Agriculture Record Keeping, Personal Finance, and Agriculture Construction
- Managed Agriculture Science classes of 10 to 30 students
- Planned social and academic events for 200 students
- Trained students for regional FFA public speaking contest

Willard High School Practicum Student
Willard R-2 School District, Willard, Missouri / January 2010-May 2010
- Observed and applied different teaching techniques
- Directed students in a laboratory and shop setting
- Advised 10 students in academic choices
- Assessed advanced student animal science projects

Ozark High School Practicum Student Teacher
Ozark R-6 School District, Ozark, Missouri / October 2009-December 2009
- Organized academic social events and delegated related activities to student committees
- Trained leadership teams of 25 students
- Responsible for 25 students during long-distance student conference

Contributed by Dr. Toree Williams. Copyright © Kendall Hunt Publishing Company.

TERRI LYNN SMITH
240 Scenic Drive, Saint Charles, Missouri 63320
terri129@missouristate.edu
310-310-1199

PROFESSIONAL EXPERIENCE

On-Site Manager of Darr School of Agriculture
Missouri State University, Springfield, Missouri / January 2010-December 2011
- Managed livestock health, nutrition and training
- Managed building facilities, barn, arenas, pastures and grounds
- Supervised student staff responsible for animal care and handling
- Hosted benefactors and visitors with scientific, academic, and government interests
- Assisted in new student outreach and large special events

Supervisor of Animal Handling & Instructor
Therapeutic Riding of the Ozarks, Springfield, Missouri / January 2009-May 2009
- Oversaw preparation and handling of horses in therapeutic riding sessions
- Co-directed approximately 20 volunteers during therapeutic riding sessions
- Prepared and evaluated detailed lesson plans weekly
- Taught disabled children riding skills and techniques to improve physical capability

On-Tour Handler for Grand Prix and Show Circuit
Raylyn International Show Jumping, Frederick, Maryland / June 2008-August 2008
- Managed care for $2,000,000 in equine stock
- Organized prize horses, vehicles, trailers and 100+ pieces of equipment for frequent transport
- Groomed animals, prepared athletes and tack for daily competitions and showing
- Communicated with clients and coaches to ensure health and safety of animals and riders

Riding Instructor and Counselor
Camp Cayuga, Honesdale, Pennsylvania / June 2007-August 2007
- Instructed equestrian riders in group and individual riding classes for children ages 5-15
- Supervised 12 young adults, acting as their primary care giver
- Directed approximately 20 children in artistic and athletic activities

TERRI LYNN SMITH
240 Scenic Drive, Saint Charles, Missouri 63320
terri129@missouristate.edu
310-310-1199

REFERENCES

Jamie Craig, DVM
Private Practice Veterinarian
Willow Ridge Equine Hospital
8991 Highway 209
Frederick, Maryland 36234
801-332-1945
willowridgeanhosp@gmail.com

Melissa Clark
Therapeutic Riding Instructor
Therapeutic Riding of the Ozarks
24809 Rangeland Road
Springfield, Missouri 65431
309-227-1001
TROzarks1@hotmail.com

Angie Simmons, PhD
Associate Professor
Darr School of Agriculture
Missouri State University
P.O. Box 98222
Springfield, Missouri 65431
309-434-9022
asimmons14@agr.misstate.edu

Resume Writing Checklist

Below is a checklist for items that should be found on a student's resume. Before beginning to update your current resume, start with your past resume or a rough draft of your resume and go through the checklist given below. If items are missing from the resume, an "x" should be placed in the blank in front of the item. Additionally, mark on your resume those items needing to be corrected and add comments at the end of this checklist to help remind you of items to work on.

Materials:

- Select white, ivory, or light-colored, high quality, cotton-fiber 8 ½" x 11" resume paper (not plain typing paper).
- Present it in a professional manner such as a plain folder with the cover letter. Avoid punching holes to put in a binder or stapling. Using a basic paperclip can be useful to keep papers together or a folder that can bind the papers together without wrinkling or punching holes.

Formatting:

- Use traditional font such as Times New Roman, Arial, or a similar font for the body of the resume. More creative font can be used to emphasize the headings or the header as long as the font is still easy to read.
- Keep to 10-12 point font for the body of the resume using larger font only for the header at the top of the page. Larger font helps your header stand out.
- Stay consistent with the margins throughout the resume and avoid margins less than ½ inch.
- Design your resume to have enough empty/open space so that the resume is easy to read and visually appealing. Avoid too much empty space that it looks like the applicant lacks experience or too little so that the resume appears cluttered and unorganized.
- Headings should stand out using bolding, capitalization, or underlining.
- Keep sections consistent in spacing and lined up.
- Format for how you word information within the body of your resume should be consistent so if, for example, you spell out a state, then spell it out throughout.
- If you use a second page, it should at least go halfway down the page, and the second page should have a similar heading and formatting as the first page.

Spelling & Grammar:

- Thoroughly check for spelling errors, typos, and grammatical errors.
- Numbers should be written out if under 10 with numerals used only for numbers 10 and up.
- Abbreviations should be avoided. Spell out titles of organizations, certifications, etc.

Education & Experience:

- Education information should include school, location, degree, GPA, and graduation or expected graduation date.
- Academic honors, awards, scholarships, etc. should be highlighted along with relevant coursework and associated projects. Project information should include title, dates, and explanation of activities.
- Skills related to industry (computer, technical, mechanical, laboratory, etc.) should be listed with related certifications, dates, and/or training locations and hours.
- Volunteer and/or internship experience is important and should be listed. Whether paid or unpaid, all types of working experience should include organization/company, position titles, locations, dates, skills/activities, etc.
- Organization/club membership and associated activities can be as relevant as a paid job, but when listed, it needs to display the level of responsibility and type of activities, other than just being a member. Information should include name of organization, positions, accomplishments, dates, etc.
- Use headings to divide up and organize experience while explaining the type of experience (Example: volunteer versus industry-related experience). Get creative on the wording used for the headings so that the experience stands out.

Presentation:

- Headers should include full, current, and correct information (name, address, phone number, and email address). Your name should stand out from the rest.
- Resumes should be customized for the job you are applying for so avoid unrelated experience and/or activities.
- Avoid repetition of experience, activities, and/or skills.
- Made sure experience, activities, schooling, etc. are presented in chronological order and avoid going back too far in years, preferably not past 7-10 years. Once you are in your junior/senior year of college, activities associated with high school are less relevant.

- Be specific on the description of activities such as quantities, amounts, hours, etc. and use creative words that make activities stand out. Make sure to use proper verb tense (past tense for past activities).
- Avoid pronouns such as "I, He, She, His, and/or Her" in the body of the resume.
- Unless relevant, recreational activities, interests, and/or hobbies should be left off of the resume. If relevant, make sure to present these activities in a more formal manner including specific organizational memberships, accomplishments, dates/hours, etc.

References:

- Leave off "references upon request" from the body of the resume.
- References should be listed on a separate page and should include full contact information (full name, title, address, phone number, and email address).
- Formatting for reference page should be similar to the rest of the resume.

Additional Comments:

_____

_____

_____

Chapter 10

Tips for Presenting Yourself Well in a Cover Letter

The cover letter is an essential part of the applicant's introduction as it gives a more personal look into the applicant's background. Although applicants will work first on developing their resume before going on to their cover letter, the potential employer will start first with the cover letter and decide if they want to review the resume.

The cover letter should complement the resume, filling in the gaps in the resume and highlighting certain areas that will "sell" the applicant to the potential employer. It's not just about the content of the letter as it also says to the potential employer that this applicant can develop a basic letter, write a clearly formed sentence, and take the time to correctly spell words, avoid typos, and use correct wording to express their thoughts. Since most undergraduates learn basic grammar in elementary and middle school, below are some basic rules to grammar that are often forgotten, but will assist in the development of a professional cover letter.

While a cover letter should be an expansion to the resume, spotlighting areas that are important for the potential position, a cover letter also tells employers whether the potential employee is familiar with the company and the position they are applying for. In fact, the name of the company and the full title of the position should be in the first paragraph along with where the student found the job announcement. From there, in the second paragraph, the student should expand on items that will show the potential employer the applicant understands all aspects of the job. Of course, the other important component to the cover letter is your contact information. Below is an example of a good cover letter developed for an animal science job.

Simple Grammar Rules

Grammatical errors are common in cover letters. These errors can say to your potential employer *you're too lazy to check for details* or *you're lacking basic language skills*. Either of these assumptions will be a disadvantage to the potential employee. Below are some comma and capitalization rules that can be applied in developing your cover letter.

Comma Rules:

- Use commas before and after words such as "however" and "therefore" if used in the middle of a sentence. *I would like to go to the party with you, however, I'm busy.*
- Use commas in a series of three or more words or phrases. *Was the stoplight green, red, or yellow?*
- Use a comma to separate two adjectives describing words. *She has a beautiful, golden Palomino.*
- Use a comma to separate parts of a date. *I graduated from high school on Monday, May 11, 2012.*
- Use a comma after an introductory phrase, word, or clause. *When we sat down, the waiter brought the menu.*
- Use a comma to set off quotations. *"I think your application is ready to submit," said Jane.*
- Use a comma after an expression. *Definitely, you can use my pen.*
- Use commas to separate the city from the state. *I was traveling to Houston, Texas, to visit a friend.*
- Use commas when connecting two independent clauses (but, and, yet, etc.). *The horse chased the calf, but the calf got away.*
- Use commas after a greeting and closing in a friendly letter. In a business letter, use a colon after the greeting and a comma after the closing.

    Examples:

    1. Friendly Greeting: *Dear Jack,*
    2. Friendly Closing: *Yours truly,*
    3. Business Greeting: *Ladies and Gentlemen:*
    4. Business Closing: *Sincerely,*

Capitalization Rules:

- Capitalize the names, initials, titles of people, particular groups, nationalities, pets, and languages.
    Example: *When Mr. Jack, president of the Elks Club, addressed the members, he spoke in English.*
- Capitalize days, months, and holidays.
    Example: *This year, Christmas Day falls on Monday, December 25.*
- Capitalize special places and specific buildings.
    Example: *One Shell Plaza is located across the street from Bayou Gardens.*
- Capitalize titles of written works.
    Example: *I just finished reading a book called The Robe.*
- Capitalize streets, cities, states, countries, and continents.
    Example: *The undergraduate student was from Oshkosh, WI, in the USA.*
- Capitalize names of products, clubs, and organizations.
    Example: *The Hawthorne Boat Club meets the first Saturday of every month.*
- Capitalize names of books, stories, poems, and songs.
    Example: *The Star Spangled Banner was written in 1814.*
- Capitalize and spell out numbers when they begin a sentence.
    Example: *Twenty year old, Jake worked in the extension internship position during the summer.*

Molly Christine Nicodemus
P.O. Box 420
Starkville, MS 38512
mcn622@msstate.edu
(662) 489-7119

May 27, 2017

Lincoln County Extension Service
2100-B Warren Street
Batesville, MS 39180

Mr. Tim Anderson:

I am writing to express my interest in securing the Extension Agent position at the Lincoln County Extension Office. I heard about this position during my internship with the Winston County Extension Office. I will be obtaining a degree in Animal and Dairy Sciences with a Production/Management concentration from Mississippi State University in August of 2017. With that being said, this job opportunity is a wonderful fit with my educational experience and career goals.

I have been involved in agriculture my entire life starting with me growing up on a dairy farm just north of Batesville, MS, and thus, I am familiar with the area and its attributes. While living on the farm, I worked with various ag-based youth groups assisting in youth development programs. Working on the farm and with the youth groups, I have developed a strong background with different livestock that was further enhanced during my time at Mississippi State University. These skills will be of use to the people of Lincoln County.

With my working experiences, I understand the importance of having great team dynamics. Teamwork was a part of my experience on the livestock judging team, and became a critical part of my daily work as an intern at the Winston County Extension Office. Through these experiences, I learned how to utilize everyone's strength for the betterment of the group and found that every person had much to contribute to a project. I enjoyed the team environment and would like to continue demonstrating to my next employer how I can be a critical part of a team.

As you can see from my letter, I am enthusiastic about being a part of the Lincoln County Extension Office as a County Extension Agent. I would appreciate the opportunity to interview with you to discuss my qualifications. I will be in the Lincoln County area in the coming month so I would be available for scheduling an interview. I can be reached using the contact information given above. I appreciate your consideration and look forward to hearing from you.

Sincerely,

Molly Nicodemus

Molly Christine Nicodemus

Cover Letter Writing Checklist

Below is a checklist for items that should be found on a student's cover letter. Before beginning to develop a cover letter for a recent job opening, start with a past cover letter or a rough draft of your cover letter and go through the checklist given below. If items are missing from the cover letter, an "x" should be placed in the blank in front of the item. Additionally, mark on your cover letter those items needing to be corrected and add comments at the end of this checklist to help remind you of items to work on.

- Use a traditional font (Examples: Times New Roman, Arial, Tahoma, or Verdana).
- Keep the font consistent throughout and avoid using less than 10 point fonts, except for the header.
- Keep margins consistent throughout the page and avoid using margins ½ inch or smaller.
- Header should be at the top of the page, stand out, and include your name, address, phone number, and email. Preferred header is one that matches your resume.
- Date followed by company address at the top of the page underneath the header.
- Body of the letter should begin with the name of the person that is doing the hiring. Avoid using To "Whom it May Concern". If you have no name, use "Hiring Manager", "Search Committee", or something similar.
- Throughout the body of the letter stay away from starting sentences with "I", especially the first sentence of the paragraph. Be creative in your sentence structuring so that it doesn't become redundant nor gives the impression of the applicant being self-centered.
- Include in the first paragraph company name, job title, and where you heard about the job opening. Make sure to use full and proper names.
- In the second paragraph, explain why you want to work for the company and how your skills match what the employer is looking for. Demonstrate that you have knowledge of the position, company, and industry associated with the position.
- In your last paragraph, thank the employer for his/her interest and suggest an interview or meeting in person.
- Keep your letter short, concise, and relevant avoiding going over one page in length.
- Sign your letter below the closing and type your full name underneath your signature.

Additional Comments:

_____

_____

_____

Chapter 11

Tips for Presenting Yourself Well in Person

Photo courtesy of Dr. Kristen Slater

Yes guys, it is time to dig out that tie, and yes ladies, it is time to iron that skirt. If your application, cover letter, and resume gets a thumbs up from your potential employer, the next step is an interview that is usually done in person, although with today's technology it may be done via Skype.

Some interviews can be very short, while others can take the day, but no matter what, first impressions are important. Although it may seem vain, first impressions include what you wear, how you smell, and how you hold yourself. Yes, even the cologne or amount of cologne can be an issue. Below is a discussion concerning body language as even your posture or how you move your hands or direct your eyes will say much about your character.

Body Language

Before you speak, your body is already talking for you giving others indicators of your emotions, feelings, intentions, thoughts, etc. Body language begins at the head and works all the way down to your feet, and while potential employers should be focusing on what you are saying, they are also getting signals from your body.

Let's start from when you walk in the door. A confident, straightforward step with the head up and eyes forward along with shoulders back says *I am ready for this interview*. When you go to shake the interviewer's hand, the hand should come out confidently and shake with a strong, but not too vigorous shake.

Eyes should be continuously looking at the interviewer showing that you are interested and focused on what they are saying, and not looking down, as if to say, I'm not confident enough for this position. Don't be afraid to smile, if appropriate, as that can help lighten the mood for both you and your interviewer. As you sit down, don't fidget, and if standing during the interview, don't lean or sway from side to side. Again, this shows nervousness.

You don't want to stand or sit too far away from your interviewer as that can portray a submissive or standoff attitude, but on the other hand, too close could be getting into someone's personal space. Good rule of thumb is you should be able to reach out without leaning to shake someone's hand if standing. Following that same rule, if you reach out with your hand and go past that person, then you're too close. What if you are sitting during the interview? A desk width is appropriate for interview spacing, even if a desk is not there. However, you shouldn't start rearranging seating during your interview so select a chair that is already set-up that allows for appropriate spacing.

The biggest question students have is what to do with my hands during the interview? Of course, start first with avoiding putting them in your pockets as that is too casual. Folding hands can portray a lack of openness, but too much hand movement can be distracting and can demonstrate nervousness. As you finish up the interview, come in as strong as you did when you entered, no matter how badly the interview may have gone, using a strong handshake followed by confident stance with your head held high.

Dressing the Part

A part of your body language is the clothing you wear. Nevertheless, dress seems to be especially hard for most ag-based undergraduates as they are use to hands-on activities requiring more rugged attire so the expectation of such students is to wear something similar to an interview. While many ag-based job interviews may require a tour on a farm, veterinary clinic, or research laboratory, you still don't want to show up in your jeans or scrubs.

You want professional attire that can still hold up in the interview so if there is a walk through a farm avoid such things as heels. Khakis and dress boots that are polished are usually the staple to most ag-based job interviews, and yes, a tie and jacket can still be added for a guy or a dressy blouse for a girl showing a degree of professionalism without lacking functionality. Do check prior to the interview whether there will be a hands-on component and the type of conditions you may be traveling through on a tour. Usually, if there is a hands-on component, it is done on a follow-up interview, but even at that time, don't show up in jeans with holes or a tank top.

Don't buy something brand new that you aren't comfortable in, and make sure that you test out your clothing selection with friends and family to get their feedback. This dress rehearsal should include everything from head to toe. The worksheet given at the end of this chapter includes a mock interview and this is a good time to do this dress rehearsal.

Interview Questions

Once you perfect your outer appearance for your interview, the next step is working on what you will say. While preparation is helpful, no matter how much you prepare, there will always be a few questions that may take you by surprise. Practice standard questions to help calm your nerves along with questions that can surprise you as employers want to see how you handle stress. Don't over rehearse answers as it needs to be natural, although well-prepared. A potential employer needs to know you took the time to prepare, but that you can handle yourself professionally in any situation, and thus, represent their company well.

Below are common questions an interviewee may encounter. There are no wrong answers. Just make sure that they are well thought out using correct grammar and expressing confidence and professionalism. Answers should get to the point, but be thorough enough to cover what the interviewer is asking.

If asked about previous work or certain individuals you have encountered, always stay positive as no potential employer wants to worry about a disgruntled employee talking negatively about their company. In the end, while you want to be thoughtful about your wording, be honest, give thorough, specific answers, and be yourself. If you don't fit into what a company wants, then it is not the job for you.

Common Questions:

- Tell me about yourself.
- Why did you leave your last job?
- What experience do you have in this field?
- What kind of salary do you need?
- Are you a team player?

- What is your philosophy towards work?
- Explain how you would be an asset to this organization.
- Why should we hire you?
- Tell me about a suggestion you have for this business.
- What is your greatest strength? Your weakness?
- Why do you think you would do well at this job?
- What are you looking for in a job?
- Tell me about your ability to work under pressure.
- Are you willing to work overtime? Nights? Weekends?
- How would you know you were successful on this job?
- Would you be willing to relocate if required?
- Describe your management style.
- What have you learned from mistakes on the job?
- If you were hiring a person for this job, what would you look for?
- How do you propose to compensate for your lack of experience?
- Describe your work ethic.
- What has been your biggest professional disappointment?
- Tell me about the most fun you have had on the job.
- Do you have any questions for me?

As you work through these questions, write down notes. Using notecards are helpful to organize your thoughts. Read over your notes prior to your interview, and this will help you feel more prepared.

After the Interview

Your work doesn't end after you finish your interview. Make sure to give a final and lasting impression through the use of a follow-up thank you note. This can be done with a personal handwritten note or via email. The note needs to be done in a professional manner and usually done no later than a week after the interview. This note needs to be sent to the hiring manager or the head of the search committee. It should include the title of the position you interviewed for. You should not only thank the individual for the interview, but demonstrate your continued interest in the position. Make sure you include at the end of the note your full name.

Mock Interview Checklist

It's time to see how you represent yourself in person, and what better way than by setting up a mock interview. This gives you a chance to see how you respond in an interview setting and get feedback from someone besides yourself. Select someone to do an interview such as a current employer, co-worker, professor, or college career center representative. Use the sample questions given in this chapter and make sure to dress as if you were applying for the type of job you would want.

Have your interviewer go through the checklist below and make sure that they sign the checklist. Below is a checklist for items that should be observed during a student's mock interview. If items are missing from the mock interview, an "x" should be placed next to the item with additional comments given at the bottom.

Mock interview date: _____

Mock interviewer: _____

- Items discussed by the interviewee were relevant to the type of position they would be applying for and were given in a concise and organized manner that would appeal to the potential employer.
- Body language and appearance including attire demonstrated professionalism and confidence. Individual appeared to fit the position they would be applying for.
- Spoke in a manner that was easy to hear, easy to follow, and kept listener's attention.
- Demonstrated enthusiasm, motivation, and/or determination for the position being interviewed for.

Additional Comments:

_____

_____

_____

Chapter 12

Where to Go from Here:
Goal Setting

Photo courtesy of Sherri Mitchell

Your career planning work isn't over yet. As you navigate through the job search and application process, you need to keep in mind what your goals are for your future. This includes both short and long term goals as this will direct you as to what jobs to apply for and accept. Do they meet your goals or will they get you sidetracked? Goals include all aspects of your life from family to work. While you may be just starting an undergraduate career, having goals will give you a light at the end of the tunnel when you start to get bogged down with schoolwork.

While setting goals should begin before you start your job search process, it's never too late to set goals, and more importantly, it is critical before you say "yes" to your lifelong career. Your first job is usually a building block to build upon for your future career, but you don't want to build this foundation in the wrong location. Goals also change as you navigate through your career planning and building process, and you're better off recognizing early that it's time to make changes than it is to stay stuck in a career path that doesn't fit your wants and needs. Re-evaluate your goals periodically and adjust accordingly. Even "dream jobs" change over time.

So where do I begin? Start with short term goals as they are the easiest to reach, and thus, can seem less complicated to achieve. Short term goals should be attainable and should be able to be achieved within a semester for an undergraduate student or at least within the school year. Think of something specific that can be measured and done within a certain amount of time. While it might not be attaining a lifelong career, it should be a goal that can help to build upon long term goals such as getting a 4.0 GPA or getting elected for a club officer position.

As you work on developing short term goals, you will also want to start establishing what you want for long term goals as the short term goals build up to long term goals. While referred to as long term, you are looking at goals that you could realistically reach within five years. If you make your long term goals too long to achieve, frustration can begin to build. Developing long term goals needs to include setting up steps to reach those goals. Most importantly, through this development process for short and long term goals you will need to write everything down as this makes the goals less abstract and more real to you. In the end, the main goal is to find yourself in the animal science career that you have dreamed of and this can be done with preparation and planning. Below are some basic tips to keep in mind as you set your goals.

Tips for Goal Setting

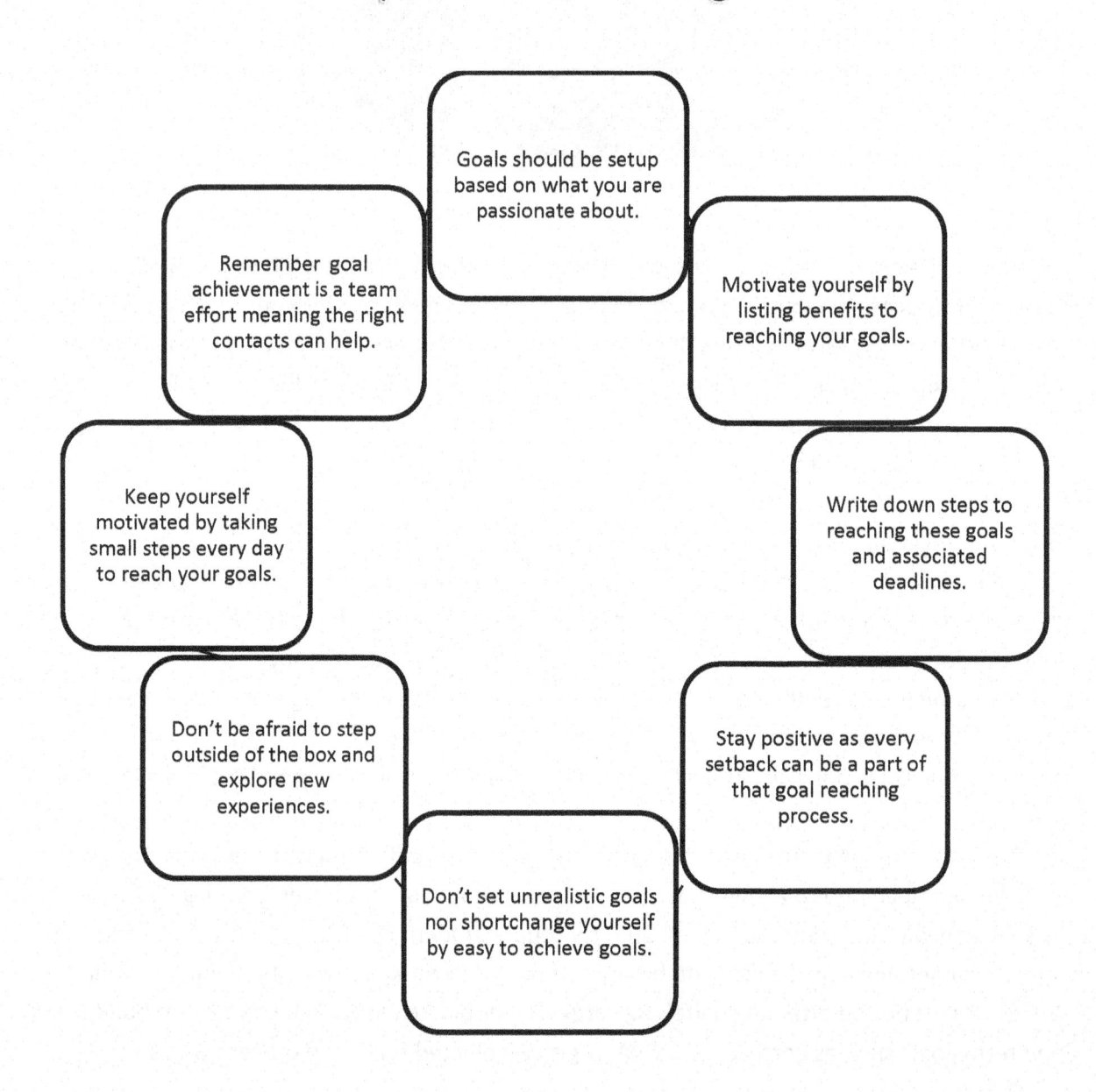

- Goals should be setup based on what you are passionate about.
- Motivate yourself by listing benefits to reaching your goals.
- Write down steps to reaching these goals and associated deadlines.
- Stay positive as every setback can be a part of that goal reaching process.
- Don't set unrealistic goals nor shortchange yourself by easy to achieve goals.
- Don't be afraid to step outside of the box and explore new experiences.
- Keep yourself motivated by taking small steps every day to reach your goals.
- Remember goal achievement is a team effort meaning the right contacts can help.

Goal Setting Activity

Hopefully, you have spent the semester thinking of your career goals as you worked to determine the career that is right for you, and while you have worked on developing an application, cover letter, and resume to achieve that career. It's now time to formalize these goals by using this worksheet to write these goals down and sharing these goals with others. This activity should make your goals real and set.

Goal setting should not end here, but be a regular part of the start of your semester followed by an assessment at the end of the semester to evaluate your successes. Keeping that in mind, consider your next semester and fill in the questions below determining your short term goals as it relates to your long term goals.

Describe your short term goal.

_____

_____

_____

Describe the steps you will do to achieve this goal (include at least three and be specific).

_____

_____

_____

How is this goal relevant to your long term goals?

_____

_____

_____

What obstacles do you expect to face in trying to reach this goal (give at least one) and how you will try to overcome this obstacle (give at least one example)?

_____

_____

_____

Give a timeline for achieving the goal. Include the end date and a mid-point where you will be able to assess your steps for reaching the goal.

_____

_____

_____